RACING CARS

RACING CARS

Written by Chris Maynard • Illustrated by Mark Bergin

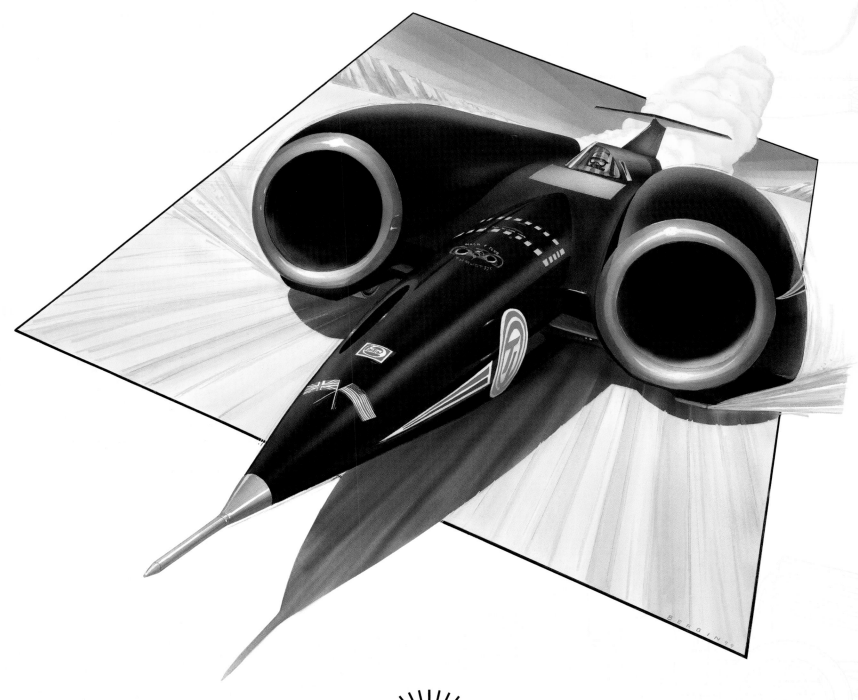

BARRON'S

First edition for the United States,
its territories, dependencies, Canada,
and the Philippine Islands, published
1999 by Barron's Educational Series, Inc.

Original edition copyright © by Franklin Watts
U.S. edition copyright © by
Barron's Educational Series, Inc.

Art Director Robert Walster
Editor-in-Chief John C. Miles
Consultant Annice Collett
The National Motor Museum, Beaulieu, England

First published in 1999 by Franklin Watts, London
96 Leonard Street, London EC2A 4XD

All inquiries should be addressed to:
Barron's Educational Series, Inc.
250 Wireless Boulevard
Hauppauge, NY 11788
http://www.barronseduc.com

Library of Congress Catalog Card No. 98-74906
International Standard Book No. 0-7641-5195-9

Printed in Hong Kong

9 8 7 6 5 4 3 2 1

CONTENTS

INTRODUCTION

THE REMARKABLE CARS IN THIS BOOK are all very different from each other. The very first one was an awkward-looking machine that won the 1906 Grand Prix motor race in France because it broke down less often

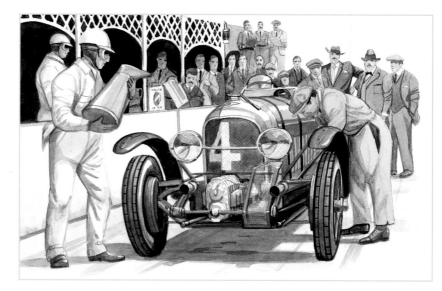

▲ In the pits at Le Mans, 1930

than its rivals. Today, racing cars are sleek and designed to shave fractions of seconds off the race times of their competitors. They are even built from high-tech materials pioneered in space programs.

▶ Driver Tazio Nuvolari wins the 1935 German Grand Prix in an Alfa Romeo P3

Despite seeming so different, all the racing cars chosen here have one thing in common—they are the very best of their time. That is how they made their mark in car-racing history. Their designers, mechanics, and drivers were pretty special, too. Without them these machines would have never performed the way that they did.

▲ Driver Juan Manuel Fangio wins the 1957 German Grand Prix in a Maserati 250F

Early racing drivers flung their cars around circuits with bravery and daring. Covered with oil and dust thrown up by the dirt tracks on which they raced, they needed lots of luck to make it to the finish line. Modern racing teams set lap times that could only be imagined by car-racing pioneers. This book contains some of the stories behind the machines, whether the race was against other cars or an earlier speed record.

▲ The revolutionary Lotus 49 of 1967

So don your helmet, fasten your safety harness and come along for a race with the very best.

1906 RENAULT

At first, Grand Prix races were more about endurance than speed. Simply finishing a race could put you in the top ten, and if you broke down less often than your opponents, you'd probably win!

◄ The chief driver for Renault was the Hungarian, Ferenc Szisz.

The circuit at Le Mans, near Paris, was the site of the first French Grand Prix in 1906. The course was 62 miles (100 km) long and the plan was to have the cars complete 12 laps over two days—a total of 750 miles (1,200 km). Each car had a crew of two. All the big car builders entered teams, but Renault had a couple of key advantages. Hydraulic dampers (shock absorbers) eased the bumps as drivers raced over poorly made roads. More importantly, quick-release rims on the rear wheels permitted fast tire changes.

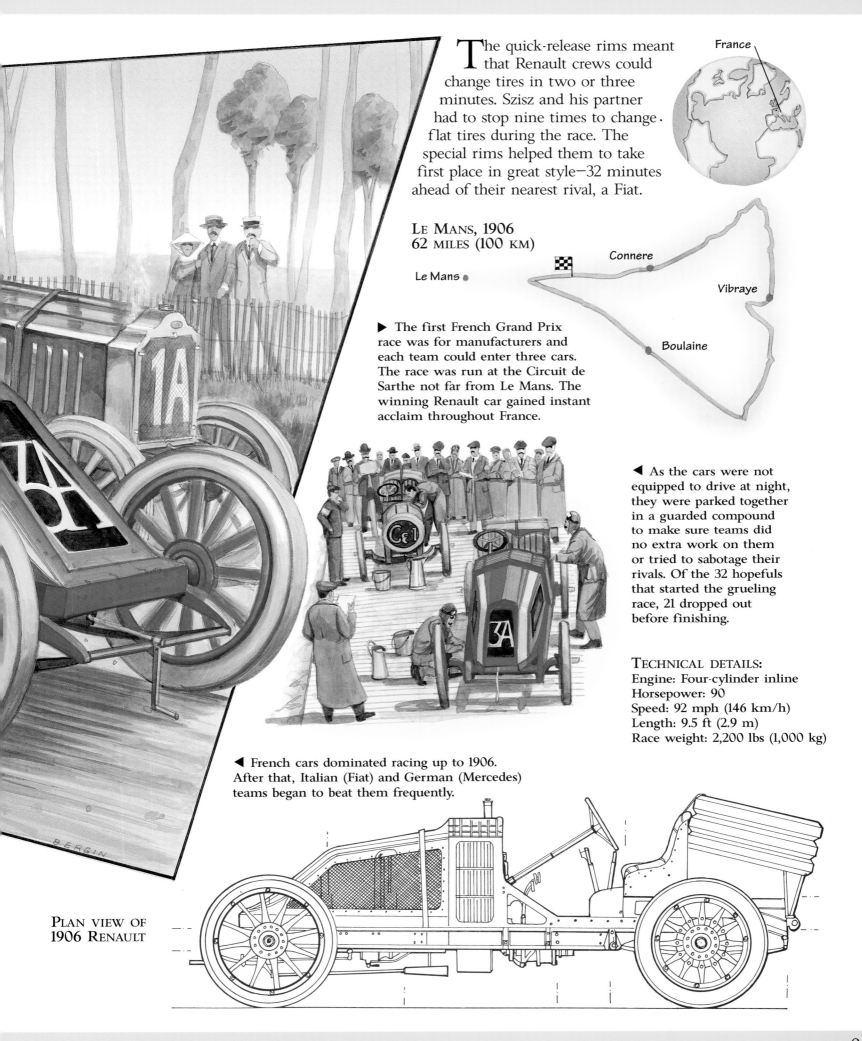

The quick-release rims meant that Renault crews could change tires in two or three minutes. Szisz and his partner had to stop nine times to change flat tires during the race. The special rims helped them to take first place in great style—32 minutes ahead of their nearest rival, a Fiat.

France

LE MANS, 1906
62 MILES (100 KM)

Le Mans

Connere

Vibraye

Boulaine

▶ The first French Grand Prix race was for manufacturers and each team could enter three cars. The race was run at the Circuit de Sarthe not far from Le Mans. The winning Renault car gained instant acclaim throughout France.

◀ As the cars were not equipped to drive at night, they were parked together in a guarded compound to make sure teams did no extra work on them or tried to sabotage their rivals. Of the 32 hopefuls that started the grueling race, 21 dropped out before finishing.

TECHNICAL DETAILS:
Engine: Four-cylinder inline
Horsepower: 90
Speed: 92 mph (146 km/h)
Length: 9.5 ft (2.9 m)
Race weight: 2,200 lbs (1,000 kg)

◀ French cars dominated racing up to 1906. After that, Italian (Fiat) and German (Mercedes) teams began to beat them frequently.

PLAN VIEW OF
1906 RENAULT

MARMON WASP

The first-ever Indy 500 race was run in May 1911. It ended in uproar. First one driver thought he won, then the title went to the driver of the yellow and black Marmon Wasp.

Ray Harroun started the race way back in the field, but a big accident halfway through allowed him to slip into the lead. However, in the confusion of wrecked cars on the track nobody realized it at the time.

Although he crossed the finish line behind another car, it wasn't until the following morning that his true winning time of 6 hours and 41 minutes became known.

▼ The Wasp took its name from its yellow and black paintwork.

▲ Ray Harroun began the race as 28th in a field of 40, yet he emerged as the winner with an average speed of a shade under 75 mph (120 km/h). The average speed today is 186 mph (298 km/h).

◀ A distance of 500 miles was the furthest cars could travel in a race and still leave the audience time to get there and go home in daylight. To add drama, a winner's prize of $25,000 was offered. It guaranteed that all the top drivers in the United States would show up.

Harroun was the only driver who had no riding mechanic to keep an eye out for other cars. He got around the problem by rigging up a rear-view mirror - possibly the first time this was ever used.

▶ The Indianapolis Motor Speedway is a 2.5-mile (4-km) long oval. For many years it had a surface of paved bricks and a nickname of "The Brickyard."

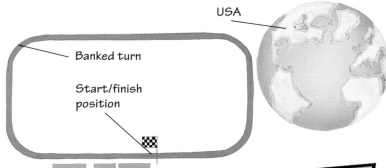

USA

Banked turn

Start/finish position

PLAN VIEW OF MARMON WASP

TECHNICAL DETAILS:
Engine: 6-cylinder
Average speed:
75 mph (120 km/h)

BLITZEN BENZ

The fastest man in America took the fastest car in Europe and, in 1910, set a new world record. It was the perfect racing partnership.

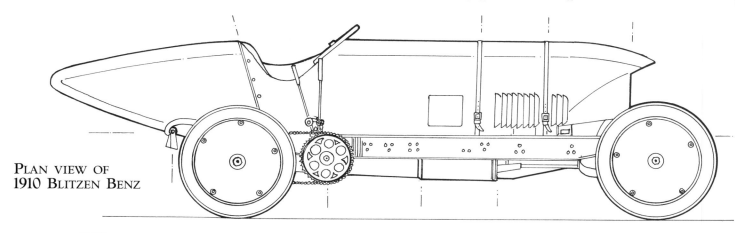

PLAN VIEW OF
1910 BLITZEN BENZ

▲ The Mercedes Benz badge—still in use on their modern cars.

Barney Oldfield was a larger-than-life racing driver and an out-and-out showman who built his career as the fastest man in the U.S.A. He would roll into town in his private railway car and challenge all takers to race at the local horse track. The crowds swarmed, the race started, and Barney thrashed his huge machines around the circuit, spattering himself with oil and throwing up clouds of dirt with his huge skidding turns.

In 1910, Barney got his hands on Victor Hemery's rocket-like Mercedes racing car, built in 1909 to attack the Land Speed Record. He named it the Blitzen (German for "lightning") Benz and took it to Daytona Beach in Florida where he set a speed record of 131 mph (210 km/h). A year later, the car's next owner, Bob Burman, pushed the monster to 141 mph (226 km/h). That record stood until 1919.

◄ At the same time as they held the world speed record, Mercedes cars had great success in Grand Prix racing. In the French Grand Prix of 1914 they crushed the opposition and took 1st, 2nd, and 3rd places. The cars' engines were based on an air engine design. Although they could belt along at 112 mph (180 km/h) this was nowhere near to what the Blitzen Benz could do.

▼ Barney Oldfield made a fortune touring the country and racing on dirt tracks against local opposition. All across the U.S.A. he was known as the "Speed King of the World."

The power from the massive ▲ engine was transferred to the rear wheels via a chain drive. Many early cars used this system.

TECHNICAL DETAILS:
Engine: Four-cylinder, 21 liter
Horsepower: 200
Speed: 141 mph (226 km/h)
Length: 16 ft (4.9 m)
Weight: 2,650 lbs (1,200 kg)

BLOWER BENTLEY

Sir Henry "Tim" Birkin was one of the leading drivers of the famous Bentley teams of 1927 to 1930 that trounced the competition in the 24-hour endurance races at Le Mans.

The "flying wings" Bentley insignia, still used on today's models.

▶ Air rushing past the cars at high speed threatened to tear the hoods from their hinges. To fix this problem, drivers fitted big leather straps to hold the hoods down.

Bentley took a fleet of big cars with them to Le Mans each year, but the 4.5 liter model became a superstar. The engine was rugged and able to give its best hour after hour. Bentley gave the car extra power by fitting a supercharger to blow the hot exhaust gases back into the intake. Everybody called the new machines "Blower Bentleys."

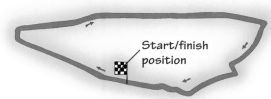

▼ The shape of the 9-mile (14.5-km) Le Mans racing circuit in the late 1920s and early 1930s.

Start/finish position

▼ In the pits at Le Mans, 1930.

Their powerful engines sent the Bentleys flying around the course at more than 100 mph (160 km/h). In 1929, Woolf Barnato and Tim Birkin led the race from start to finish. In 1930, Barnato and Glen Kidston averaged 76 mph (122 km/h) over the whole 24 hours, including pit stops.

PLAN VIEW OF BLOWER BENTLEY

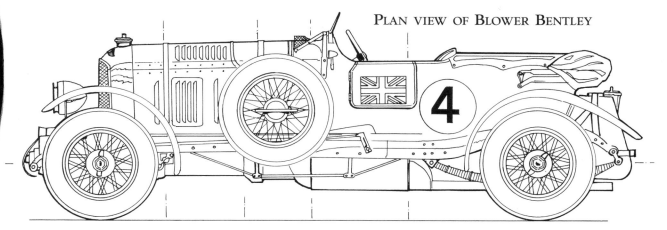

◄ Drivers in the 1920s and 1930s wore flying goggles and coveralls. While Kidston flung the big machine around the course for all it was worth, Barnato fed and rested for a few hours. Then they swapped places.

TECHNICAL DETAILS:
Engine: Four-cylinder 4,398 cc
Horsepower: 240
Maximum speed: 138 mph (220 km/h)
Length: 14.6 ft (4.4 m)

BUGATTI TYPE 35

◄ The Bugatti insignia

M any racing fans swear the Type 35 was the most beautiful racing car ever built. Though the argument can never be settled, the cars were masterpieces of workmanship and a joy to drive.

In 1924, at the French Grand Prix, the racing world had a first good look at the Type 35. With their neat good looks and gorgeous bright blue paintwork, they won over everyone. Drivers found them small, light, and a dream to handle. But for their first outing, the cars were not supercharged and were hindered by poor tires. By 1926, a new Type 35 with a 2.3-liter, supercharged engine won 12 Grand Prix races to take the world championship. A year later, it swept every honor in sight.

PLAN VIEW OF BUGATTI TYPE 35

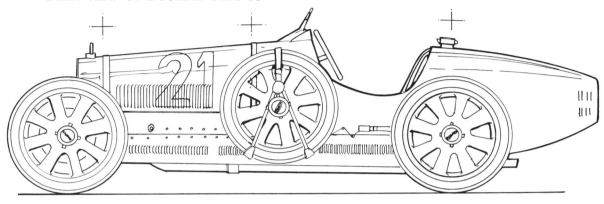

▲ The wonderfully balanced body design made for superb steering and road holding. To streamline the car, the frame was completely covered and the nose made extra narrow by putting the handbrake and gear-stick levers outside the body.

TECHNICAL DETAILS:
Engine: Eight-cylinder 2.3-liter
Horsepower: 135
Maximum speed: 120 mph
(193 km/h)
Length: 7.8 ft (2.4 m)

◀ The car was such a sensation that not until 1932, the fourth year of the race, did a car other than a Bugatti win at Monaco.

On a warm spring Sunday in April 1929, 16 cars lined up at the starter's flag for the first-ever Monaco Grand Prix. Eight were Bugattis. The circuit wound through the heart of the town and contained sharp bends and steep hills. Any driver who made an error in the narrow streets was sure to hit something. It was a fast and dangerous race - and the crowd loved it. An English driver, William Grover-Williams, won in a Bugatti with a time of 3 hours 56 minutes and 11 seconds. Of the eight finishers, five were Type 35s.

▶ A group of cars, led by a Bugatti, turn into a sharp bend in the 1929 Monaco Grand Prix.

ALFA ROMEO P3

▶ The Alfa Romeo insignia

In its first season, in 1932, the P3 won six Grand Prix races. But its most famous win of all came three years later in its career, against cars that were much faster and more powerful.

The 1935 German Grand Prix should have been a cakewalk for Mercedes or Auto Union cars. They were technically far superior to the aging P3. In the first half, the race dwindled to nine German cars and one Italian Alfa - which at this point was in the lead. At half-distance the Alfa driver, Nuvolari, pulled into the pit. But refueling and tire changes were so slow he started off again in sixth place. Driving like mad on the wet track he closed the gap until he was in second place a good minute behind the leader, von Brauchitsch, in a Mercedes W25. Then fate stepped in. The W25's left rear tire began to come apart.

If von Brauchitsch had stopped in the pit for a new tire, Nuvolari would have stolen the race. His only hope was to plunge into the last lap and hope the tire would make it. With only 5 miles (8 km) to go, the tire burst in a storm of whirring fibers. Nuvolari stormed past him and into first place. It was the old P3's most famous victory ever, and over a stronger and faster rival.

◀ For 30 years, Tazio Nuvolari astonished the racing world. He first raced cars at the age of 32 and over the years won many victories for Alfa Romeo. He was extremely daring and often pushed his car to the very edge of what it could do - yet always managed to stay in control.

TECHNICAL DETAILS:
Engine: Eight-cylinder 2.5-liter
Horsepower: 215
Maximum speed: 140 mph (225 km/h)
Wheelbase: 8.5 ft (2.6 m)

▼ The Alfa Romeo P3 was the first true single-seat racing car. Before that, all cars had an extra seat for a mechanic who helped out with problems in the days when cars were fast but not that reliable.

PLAN VIEW OF ALFA ROMEO P3

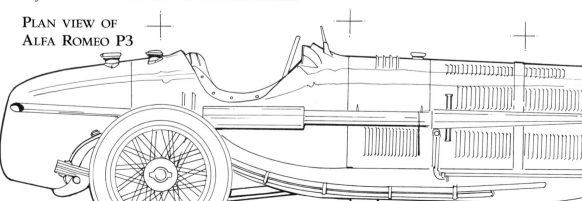

▶ As von Brauchitsch struggles to control his car, Nuvolari powers past to slip into first place and win.

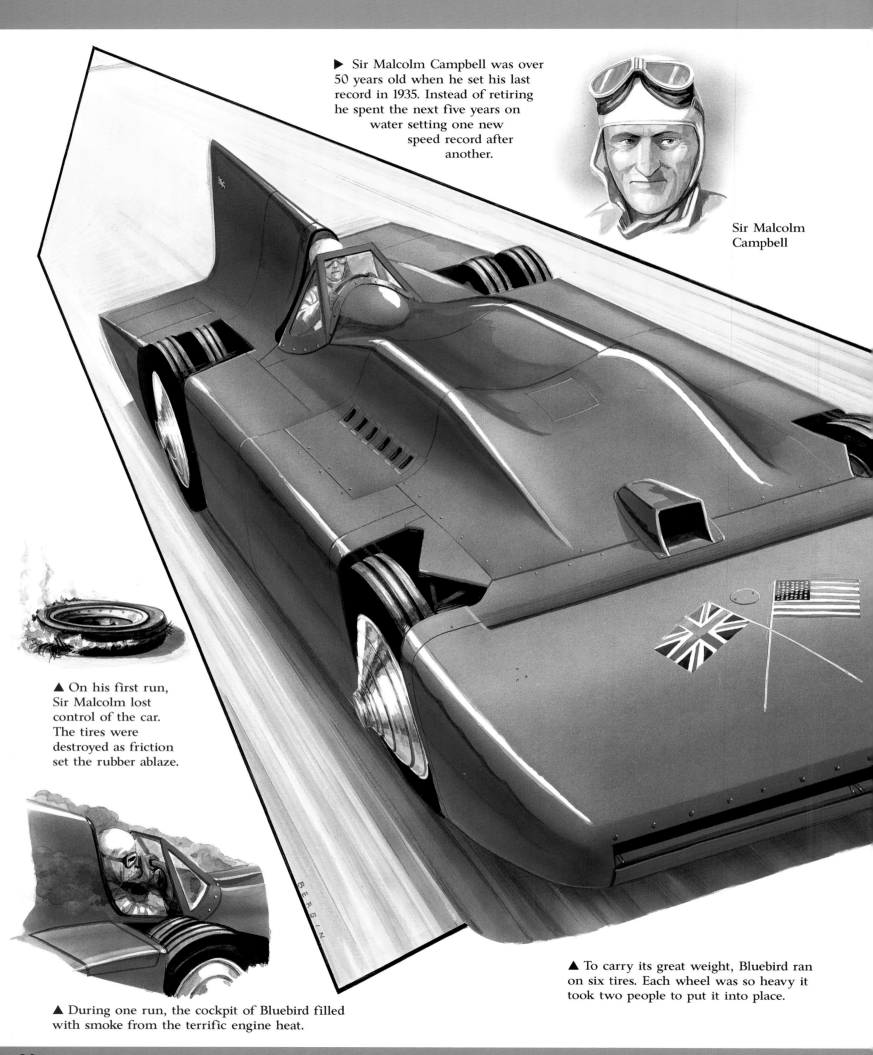

▶ Sir Malcolm Campbell was over 50 years old when he set his last record in 1935. Instead of retiring he spent the next five years on water setting one new speed record after another.

Sir Malcolm Campbell

▲ On his first run, Sir Malcolm lost control of the car. The tires were destroyed as friction set the rubber ablaze.

▲ To carry its great weight, Bluebird ran on six tires. Each wheel was so heavy it took two people to put it into place.

▲ During one run, the cockpit of Bluebird filled with smoke from the terrific engine heat.

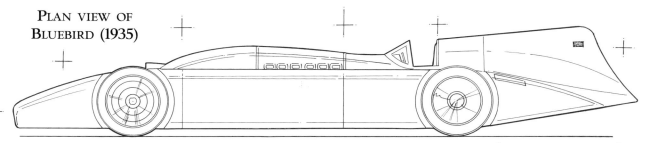

TECHNICAL DETAILS:
Engine: 36.5-liter
Rolls-Royce air engine
Horsepower: 2,500
Maximum speed:
301 mph (482 km/h)
Length: 28 ft (8.5 m)

BLUEBIRD (1935)

▼ Bluebird was as streamlined as an aircraft. To keep it hugging the ground, it had a drooping nose and skirts that covered the sides of the body.

Each car that Sir Malcolm Campbell set a speed record in was called Bluebird, after a play he saw in 1912. It must have been a great night at the theater!

The problem of going superfast is finding a straight stretch of track long enough to get up speed—and to brake again. Once drivers ran out of room on racing tracks, they took to using hard-packed sandy beaches that were level and smooth. Daytona Beach in Florida was the favorite for many years as drivers pushed beyond 200 mph (320 km/h). By 1935, Sir Malcolm Campbell had outgrown Daytona. The only other place anyone knew of that was big enough for a land speed record was in Utah, on the Bonneville Salt Flats.

In September, 1935, Sir Malcolm arrived there with a car he believed could break the 300 mph barrier. It was a monster. The 1935 Bluebird had a wide streamlined body that all but hid its wheels and a tail fin to keep it heading straight ahead. Under the hood was a huge Rolls-Royce aircraft engine. On September 3, Sir Malcolm took the car to 301 mph (484 km/h) and became the first person ever to travel that fast on land.

▲ The first Bluebird to become famous set a Land Speed Record of a shade more than 146 mph (235 km/h) in 1924. With a big 12-cylinder engine to power it, Campbell reached this speed on the flat Pendine Sands on the coast of Wales.

◄ Daytona Beach in Florida is a ribbon of hard sand, 25 miles (40 km) long. For many years in the 1920s and 1930s, it was the best course for setting speed records. The course itself was 12 miles (20 km) long with a timed section in the middle. The rest was used to get up speed and to slow down again.

◄ At Daytona Beach in 1932, in a huge new Bluebird with a tail fin and an aircraft engine, Sir Malcolm set a new record of a shade under 254 mph (406 km/h).

MASERATI 250F

◀ The Maserati had a weak rear end, so it was a bad idea to race with a full load of fuel.

Maserati was a small Italian company that didn't have deep enough pockets to stay in Grand Prix racing for long. Yet its cars were stylish and streamlined. Everybody who drove the 250F found that it steered beautifully. This car was the last Maserati to win a Formula One Grand Prix race.

▲ Even though his car had less power, Fangio stole past two big Ferraris to win the German Grand Prix by four seconds.

He trained as a mechanic, so by the time he began to drive Fangio really knew what racing cars could and couldn't do. When he retired in 1958, Fangio had won 24 races in 51 starts—an amazing feat.

Juan Manuel Fangio. ▶

In the 1950s Juan Manuel Fangio was world champion five times—a record nobody has ever beaten. But his best race was the day he won the 1957 German Grand Prix in a Maserati 250F.

The Maserati steered beautifully, but unfortunately it was underpowered compared to its rivals. Fangio knew he would have to refuel during the race, so he planned to build up a big lead before stopping. But everything went wrong in the pits and he was 28 seconds behind by the time he got back on the track.

▶ Fangio overtakes at the end of the race.

Lap after lap Fangio whittled away at the lead. Finally there was only a Ferrari in front of him. On the very last lap, he waited until the car slowed going into a curve. Then instead of braking, Fangio zipped past, trusting his car to steer itself through without losing control. It worked and he beat the Ferrari to the finish.

TECHNICAL DETAILS:
Engine: six cylinder 2,500 cc
Horsepower: 270
Maximum speed: 180 mph (300 km/h)
Length: 13.5 ft (4.1 m)

PLAN VIEW OF MASERATI 250F

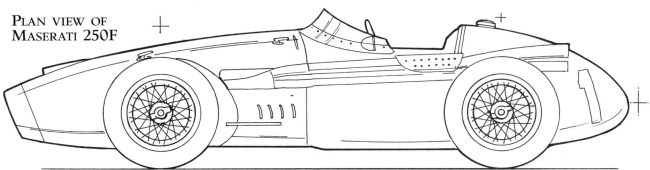

JAGUAR D-TYPE

In the 1950s, winning Le Mans meant making headlines everywhere. It was exactly the publicity Jaguar needed in order to sell its cars.

▲ After an advertising agency came up with a list of wild cats, Sir William Lyons, the company's founder, chose "Jaguar" - and so a famous name was born.

Mike Hawthorn's D-type was fitted with one of the most reliable race engines ever built and could keep up a scorching pace hour after hour. Lap by lap, for 24 hours, the Jaguar stayed firmly in the lead. The car went on to win with an average speed of 107 mph (171 km/h) and a total race distance of 2,570 miles (4,112 km).

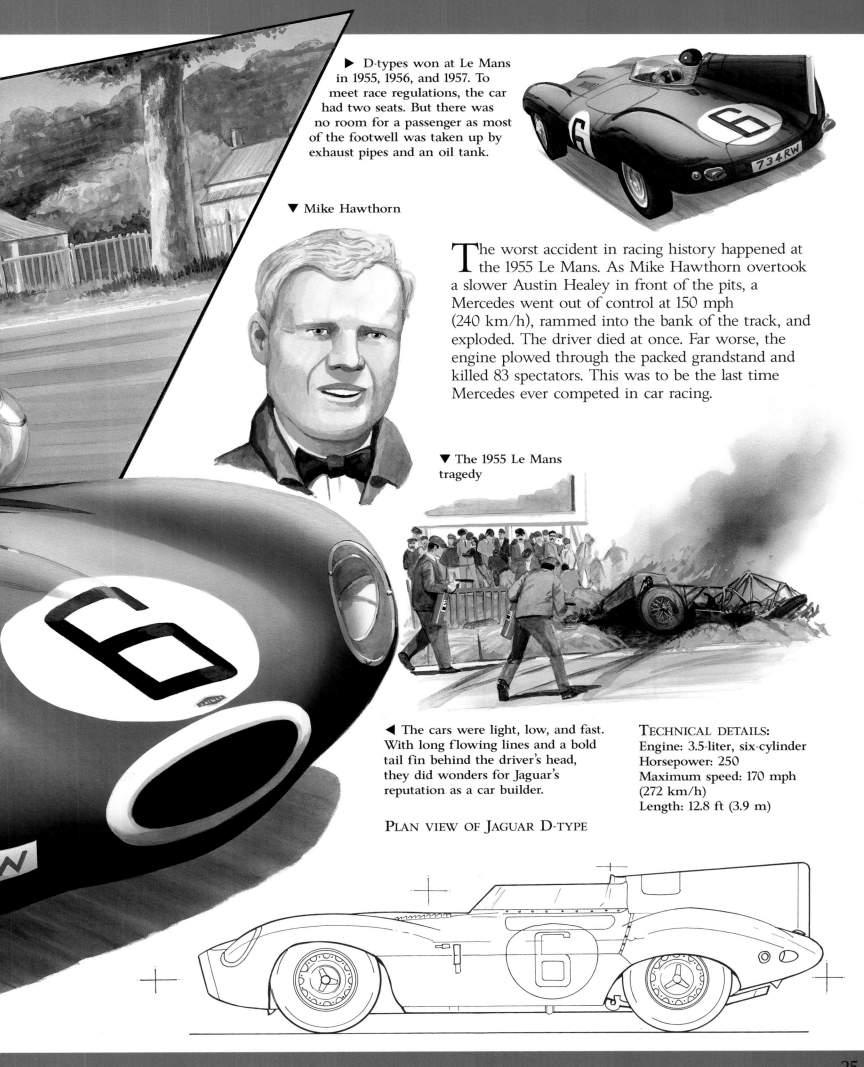

▶ D-types won at Le Mans in 1955, 1956, and 1957. To meet race regulations, the car had two seats. But there was no room for a passenger as most of the footwell was taken up by exhaust pipes and an oil tank.

▼ Mike Hawthorn

The worst accident in racing history happened at the 1955 Le Mans. As Mike Hawthorn overtook a slower Austin Healey in front of the pits, a Mercedes went out of control at 150 mph (240 km/h), rammed into the bank of the track, and exploded. The driver died at once. Far worse, the engine plowed through the packed grandstand and killed 83 spectators. This was to be the last time Mercedes ever competed in car racing.

▼ The 1955 Le Mans tragedy

◀ The cars were light, low, and fast. With long flowing lines and a bold tail fin behind the driver's head, they did wonders for Jaguar's reputation as a car builder.

TECHNICAL DETAILS:
Engine: 3.5-liter, six-cylinder
Horsepower: 250
Maximum speed: 170 mph (272 km/h)
Length: 12.8 ft (3.9 m)

PLAN VIEW OF JAGUAR D-TYPE

LOTUS 49

▶ Behind the driver's cockpit, the shell of the body stopped—everything else was open to the air. But what was really unusual about the car was the way the engine held the rear suspension and wheels in place. The engine was the body!

◀ Colin Chapman was a master builder of Grand Prix cars and the founder of Lotus Racing.

L otus came up with a completely original racing car in which a rigid engine was bolted between the body and the rear wheels. It was a huge success.

T he first race appearance of the Lotus 49 was in the Dutch Grand Prix in June 1967. The car and engine were both brand new and untried in racing conditions. Two cars were entered, driven by Jimmy Clark and Graham Hill. Hill set the fastest times in the practice laps. In the race itself his Lotus took the lead right away and held it until

the camshaft gear broke in the eleventh lap. Clark then took over from Hill and went on to lead the pack all the way to the checkered flag at the end. It was an astonishing run of luck—new engines usually have teething problems—and one of those rare moments in racing when a brand new car goes out and drives off with the prize.

▲ Lotus Engineering build racing cars and sports cars.

Some say Jimmy Clark was the most naturally talented driver of all time. As long as his cars didn't break down during a race, chances were he would win it. As one of the main drivers for Lotus, he racked up a string of successes. In eight seasons of racing and just 72 starts, he won 25 races, finishing second only once. He was killed in a racing crash in 1968, at the age of 32.

▶ Jimmy Clark

▶ From 1968 to 1976, the Ford Cosworth DFV engine was used by almost every world-champion driver.

Lotus leaped to the top of Grand Prix racing with an engine that was built by Cosworth and paid for by Ford. The Ford Cosworth DFV (the initials stand for "double, four-valve") made racing history. It became the most successful engine ever, with 174 wins.

PLAN VIEW OF LOTUS 49

TECHNICAL DETAILS:
Engine: 3 liter V-8; Horsepower: 415
Maximum speed: 149 mph (239 km/h)
Length: 13.1 ft (4 m)

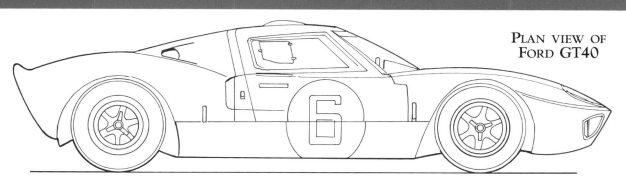

FORD GT40

In the early 1960s, Ford decided to try its hand at long-distance racing. It built the amazing GT40, and then went out and trounced the competition.

At Le Mans in 1965, the GT40s got off to a flying start. Though they led the pack from the outset, after eight grueling hours of roaring around the circuit all the Fords had dropped out. Their engines were powerful and reliable, but the rest of the car just wasn't up to the 24-hour pounding.

After this setback, Ford engineers went back to the drawing board. By shortening the nose of the car and beefing up the radiators and brakes, they boosted the stamina of the cars. In the 1966 race, GT40s dominated the pack and came in first, second, and third. In 1967, the cars came back and did it again —and in 1968 and 1969 too!

◀ Fresh wheels were bolted on seconds after the car rolled to a stop. The pit crew pumped fuel through high-pressure hoses while a fresh driver leaped in to keep the car racing around the clock.

LE MANS, 1966-69

Tertre Rouge corner

Mulsanne straight

Arnage corner

Start line/pits

▶ The track at Le Mans was more than 8 miles (12 km) long, yet the big-engined cars that raced here could get around it in under four and a half minutes. The GT40, with its huge V8 engine, could accelerate to over 200 mph (320 km/h) in the long Mulsanne straight.

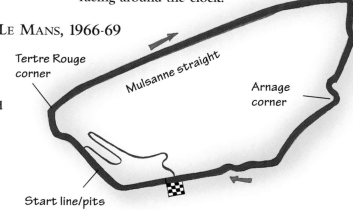

▶ The Ford insignia

▼ Jacky Ickx drove a GT40 (number 6) to the closest victory ever seen at Le Mans in 1969. After 24 hours of racing, he beat the competition by a mere 0.35 mile (0.12 km).

◀ The GT40 took its name from its low 40-inch roof.

▲ The same GT40 car that won the race in 1968 also won in 1969—a first for Le Mans.

TECHNICAL DETAILS:
Engine: 5-liter, rear-mounted 8-cylinder
Horsepower: 425

Maximum speed: about 217 mph (347 km/h)
Length: 14 ft (4.2 m)

TOP-FUEL DRAGSTER

Drag racing started in the 1950s on empty roads and unused airport runways. By the 1970s, it had its own courses, trophies, and officials. Thousands came to watch big races.

Top-fuel dragsters race singly or in pairs on a quarter-mile (400-meter) straight track. By the 1970s, the biggest and fastest machines had a narrow body with tiny, tricycle-sized wheels up front. In back were two overweight rolls of rubber that looked like they had been borrowed from a 747. Meanwhile the driver squeezed into an impossibly tiny cockpit in front of an engine so big it looked like it was spilling out of the car.

▶ A parachute deploys to slow a dragster down at the end of its run.

▶ Drag racing is about showmanship as well as power.

The huge gleaming engine usually had eight cylinders and it ran on a strange fuel cocktail called nitro-methanol. When this technology was put in the hands of a really skilled driver, like Don Garlits, the result was almost unbelievable. In 1975, for example, he set a quarter-mile (400-meter) record of 250 mph (400 km/h) in 5.63 seconds—a lot faster than it takes to read this sentence.

▼ Starting signals are rigged up on a "Christmas tree." The lights tell drivers when to roll forward into position, when the countdown has started, when to start racing, and if they are disqualified for starting too soon.

◀ To make their big rear slick tires grip better, drivers burn them in before a race. They splash water over the tires, lock their front brakes, and rev hard. The rear wheels spin, gushing smoke and flames. As the rubber softens it forms a sticky grip with the track that lets drivers accelerate hard without slipping or wasting power.

◀ There are two main ways to boost dragster engine power: with a supercharger that blows air and fuel into the cylinders and with a fuel called nitro-methanol that burns more violently than gasoline.

TECHNICAL DETAILS:
Engine: 8.4-liter eight-cylinder
Horsepower: 2,000+
Maximum speed:
250 mph (400 km/h)
Body: Steel frame and fiberglass shell

PLAN VIEW OF
TOP-FUEL DRAGSTER

INDY CAR

In 1992, Al Unser Jr. amazed spectators at the Indy 500 race by bringing his car over the finish line just four-hundredths of a second ahead of the competition. That's less time than the blink of an eye!

Unser won at the Indianapolis Speedway for the first time in his career (he won again in 1994) driving a Galmer 92 Chevrolet. After a shade over three hours of hard driving he crossed the finish line first, inches ahead of his rivals. His car had a turbocharged Chevy V8 engine in back. It carried big airfoils front and rear that forced the wheels down at high speeds and improved their grip on the track immensely. The rear spoiler, which only weighs about 20 lbs (9 kg), created a downforce at racing speeds of about 4,500 lbs (2,040 kg)—three times the weight of the car.

▼ The pits at Indy 500 races are scenes of hectic activity as crews refuel, change tires and make any repairs needed—all in double-quick time to get the car and driver back on the track as soon as possible.

▶ For races on oval circuits, Indy cars run on staggered wheels. This means the outer tires are slightly taller than the inner tires. This gives more control through the tight turns.

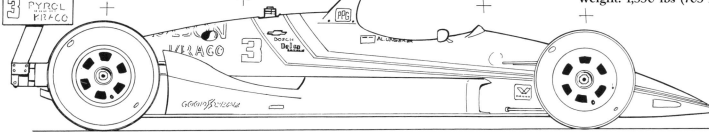

PLAN VIEW
OF INDY CAR

TECHNICAL DETAILS:
Engine: 2.6-liter, eight-cylinder
Horsepower: 720
Maximum speed: 230 mph (368 km/h)
Length: 15.4 ft (4.7 m)
Height: 32 in (81 cm)
Weight: 1,550 lbs (705 kg)

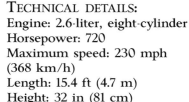

▼ Al Unser is one of the best drivers in the sport today with over 30 Indy car victories to his name, including two wins at the Indy 500.

Sitting in an open cockpit at 200 mph (320 km/h) puts a lot of strain on the head. The curve of the helmet top is like a wing and it tries to lift the head that's inside it. Helmets are fitted with slots and spoilers to break up the air flowing over them and ease the lifting effect. Even more tiring is the way the force of gravity multiplies 2.5 to 3 times as cars take tight turns. It makes an 180-lb (82-kg) man feel like 540 lbs (245 kg). Drivers need to be fit to stand hours of this sort of strain.

◄ The 76th Indy 500 in May 1992, saw more than 450,000 paying spectators at the track. The Indy 500 draws the largest-paying crowds of any sporting event in the world.

THRUST SSC

On October 15, 1997, Thrust SSC (SuperSonic Car) set a first supersonic land speed record of Mach 1.02. This was 100 mph (160km/h) faster than the existing record.

Richard Noble ▶ first set a Land Speed Record of 633 mph (1,018 km/h) in 1983 with his car Thrust 2.

As the Thrust team looked on, the huge black car ran silently past, followed moments later by a twin "boom" as sound struggled to keep up. The run began at low power to stop the two jet engines from sucking in desert dirt. Once the car was rolling fast enough to get plenty of air the driver throttled up to full power. Next he switched to minimum afterburner, then maximum afterburner before passing the timing lights at Mile 6. Five seconds later he started to slow.

At 600 mph (965 km/h), the braking parachute deployed. Two minutes after starting, the 13-mile (21-km) run was over.

▲ Thrust 2, which Noble used to claim the speed record in 1983.

◀ Of the dozens of applicants for the job of driver, Noble chose Andy Green for one simple reason—he knew about going supersonic. Green was a Royal Air Force pilot, trained to fly Tornado aircraft. His other hobbies, bobsleds and motorbikes, showed speed held no fears for him. At home, he drove a Toyota.

Thrust SSC's solid ▶ aluminum wheels weigh 300 lbs (136 kg) each. At full speed they turn at 8,000 revolutions per minute.

▶ The team refuels Thrust SSC for its second run.

TECHNICAL DETAILS:
Engine: Two Rolls-Royce Spey jet engines
Speed: Mach 1.02 (763 mph or 1,220 km/h)
Length: 54 ft (16.5 m)
Width: 12 ft (3.7 m)
Weight: 8.5 tons (8.7 tonnes)

PLAN VIEW OF THRUST SSC

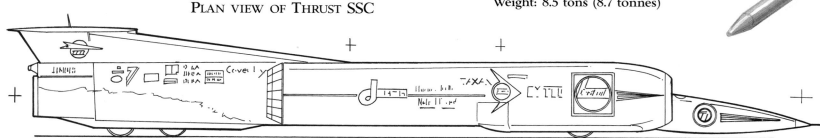

▼ In Reno, Nevada, the car got a souvenir speeding ticket from the police for doing 700 mph in a posted 55 mph zone!

▲ At the end of its run, the car was refueled to prepare for the return leg. World record rules require two runs in opposite directions, made within one hour of each other. The runs took place in the Black Rock Desert, 105 miles (170 km) north of Reno, Nevada, U.S.A. The desert provided the team with a hard, dead-smooth surface with no stones. At 15 miles (24 km) long and 5 miles (8 km) wide, its size was perfect too.

PORSCHE GT1

As a fitting way to celebrate the company's 50th birthday, the cars from the Porsche team came in first and second in the 1998 endurance race at Le Mans—an amazing victory!

◀ Ferdinand Porsche started the company in 1948. He soon guessed that if his cars won races it would do wonders for sales. His hunch was right.

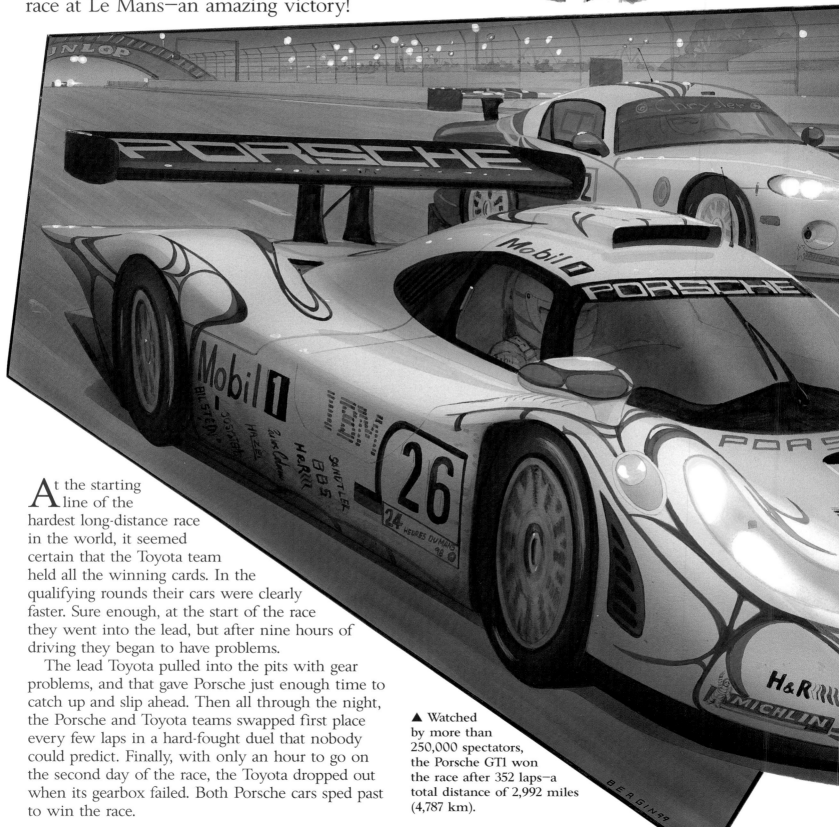

At the starting line of the hardest long-distance race in the world, it seemed certain that the Toyota team held all the winning cards. In the qualifying rounds their cars were clearly faster. Sure enough, at the start of the race they went into the lead, but after nine hours of driving they began to have problems.

The lead Toyota pulled into the pits with gear problems, and that gave Porsche just enough time to catch up and slip ahead. Then all through the night, the Porsche and Toyota teams swapped first place every few laps in a hard-fought duel that nobody could predict. Finally, with only an hour to go on the second day of the race, the Toyota dropped out when its gearbox failed. Both Porsche cars sped past to win the race.

▲ Watched by more than 250,000 spectators, the Porsche GT1 won the race after 352 laps—a total distance of 2,992 miles (4,787 km).

▲ The GT1 was the 16th Porsche to win at Le Mans - making the manufacturer more successful in this long-distance race than any other.

PLAN VIEW OF
PORSCHE GT1

▼ Inside the Porsche 917 was a huge 12-cylinder engine that made it the fastest car in racing—it did 236 mph (380 km/h) at Le Mans in 1969. In 1970, it took 1st, 2nd, and 3rd place. In 1971, it won every race it entered. By 1974, the 917 was finally banned from competition—there was nothing to beat it!

TECHNICAL DETAILS:
Engine: 6-cylinder 3.2-liter turbocharged
Horsepower: 550
Maximum speed:
220 mph+ (350 km/h+)

Of the 73 cars that started the 1998 race, only 23 finished. The rest dropped out as their cars broke down or had accidents during the long hours of being hammered around the course. The lap speed record for the race was just over 137 mph (220 km/h).

◀ The Porsche 911 road car is stuffed with bits and pieces first tried out in racing: from the streamlined body shell to the fuel management and anti-lock brake systems (ABS).

FERRARI F 300

Every year, Ferrari rolls out a new Formula 1 car for the next racing season. From the moment people saw the F 300 they knew it was a winner.

◀ Enzo Ferrari (1898-1988). The first time a Ferrari competed in a Grand Prix race was in May 1947.

▶ Schumacher storms to victory at the 1998 British Grand Prix.

The F 300 that greeted the world in January 1998 was completely new from the inside out: new engine, new gearbox and new body with loads of safety features to protect drivers in case of accidents. But modern race cars are not just built and pushed out on the track. Changes are made every day. Engineers fiddle with the foils. Different tire sets are tried out. The electronics and computers are tweaked. It's all done with an eye to wringing every last bit of performance from the car.

Despite the awful driving conditions of the 1998 British Grand Prix, driver Michael Schumacher showed his rivals what everyone on the Ferrari team already knew—they had a race-winning car.

▲ The Ferrari team tested the F 300 in a wind tunnel to find the shape that would slip through the air most easily. As a result, the car had a long narrow snout and a hi-tech nose that looked like it was lifted from a hydrofoil.

Over the season the nose design changed five times as engineers fiddled to get the right balance of forces. Each improvement helped the car hug the track a bit more, so it could go just a little bit faster. With race times measured in hundredths of a second, this makes all the difference between winning and losing.

◄ Schumacher is a master of bad weather driving—a skill that helped him win the race at Silverstone. He started the course with dry tires, but once he changed to rain tires his advantage over the competition was never in doubt. His Ferrari teammate, Eddie Irvine, came third in this race.

◄ In 1998, Michael Schumacher won six races in all and came second in the Driver's World Championship.

TECHNICAL DETAILS:
Engine: 3-liter, 10-cylinder engine
Horsepower: 700

Speed: 122 mph (195 km/h)
Length: 14 ft (4.3 m)
Weight: 1,320 lbs (600 kg)

PLAN VIEW OF FERRARI F 300

BUILDING F1 CARS

Building a Formula 1 car isn't only about speed. The cars have to protect their drivers and be reliable as well. Running fast but breaking down a lot won't win any Grand Prix races.

Formula 1 cars are born in small hi-tech factories that have nothing in common at all with the ranks of robots that make the cars the rest of the world drives. The whole process of designing and building a car is surprisingly quick. Sometime around August or September of each year, teams get to work on the new model they will launch the next January.

▲ The new car starts as a set of sketches drawn up by skilled draftsmen.

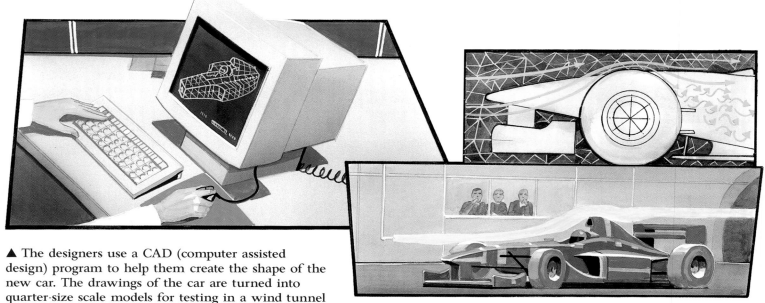

▲ The designers use a CAD (computer assisted design) program to help them create the shape of the new car. The drawings of the car are turned into quarter-size scale models for testing in a wind tunnel (right). Cars are so quick these days that winning comes down to small differences in the way they slip through the air.

▼ After the shape is decided upon, the most important job is building the body. A molded shell is made out of carbon fiber (inset), a material that's as tough as steel but with a fraction of its weight. The fresh mold is cooked for several hours in an oven until it has set hard.

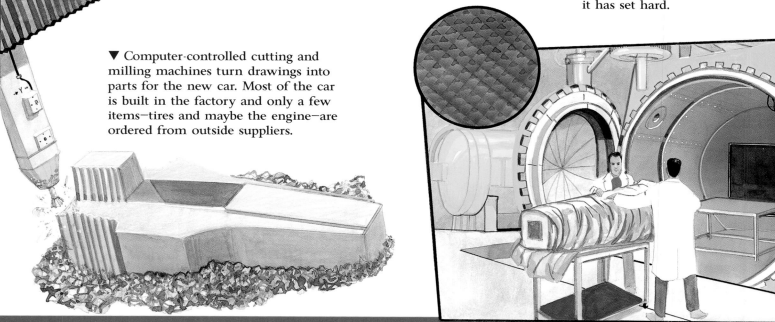

▼ Computer-controlled cutting and milling machines turn drawings into parts for the new car. Most of the car is built in the factory and only a few items—tires and maybe the engine—are ordered from outside suppliers.

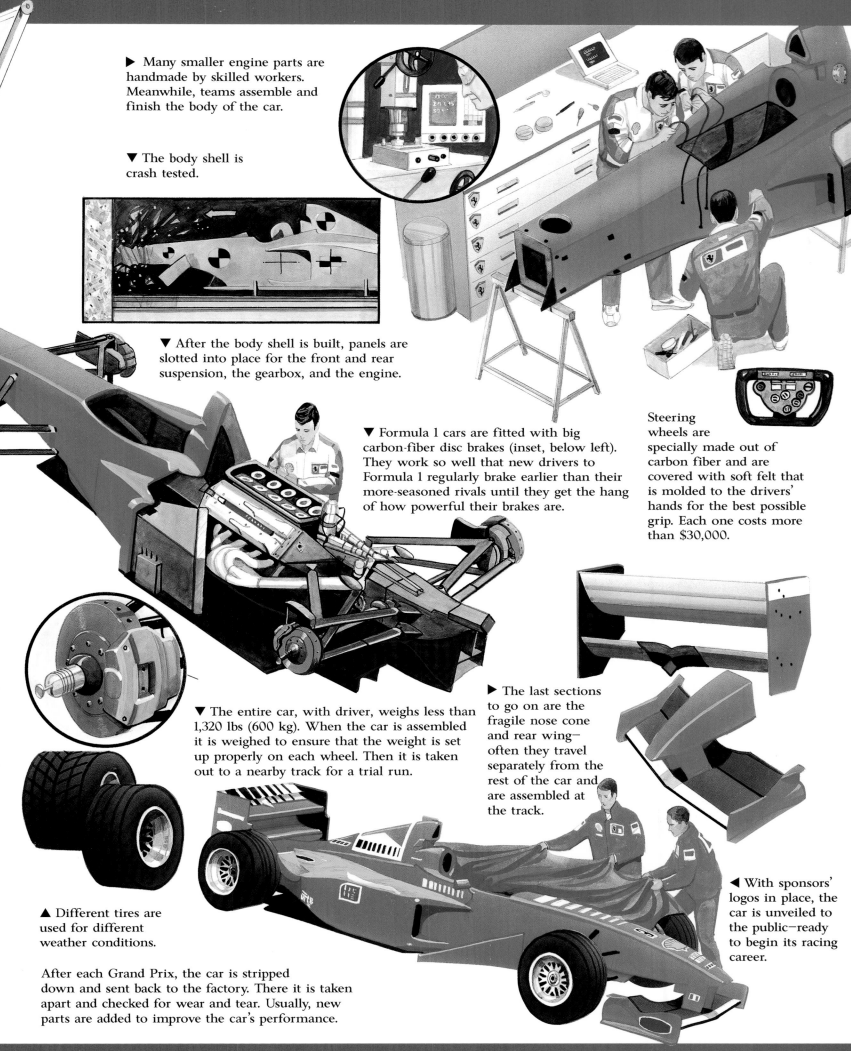

▶ Many smaller engine parts are handmade by skilled workers. Meanwhile, teams assemble and finish the body of the car.

▼ The body shell is crash tested.

▼ After the body shell is built, panels are slotted into place for the front and rear suspension, the gearbox, and the engine.

▼ Formula 1 cars are fitted with big carbon-fiber disc brakes (inset, below left). They work so well that new drivers to Formula 1 regularly brake earlier than their more-seasoned rivals until they get the hang of how powerful their brakes are.

Steering wheels are specially made out of carbon fiber and are covered with soft felt that is molded to the drivers' hands for the best possible grip. Each one costs more than $30,000.

▼ The entire car, with driver, weighs less than 1,320 lbs (600 kg). When the car is assembled it is weighed to ensure that the weight is set up properly on each wheel. Then it is taken out to a nearby track for a trial run.

▶ The last sections to go on are the fragile nose cone and rear wing— often they travel separately from the rest of the car and are assembled at the track.

▲ Different tires are used for different weather conditions.

◀ With sponsors' logos in place, the car is unveiled to the public—ready to begin its racing career.

After each Grand Prix, the car is stripped down and sent back to the factory. There it is taken apart and checked for wear and tear. Usually, new parts are added to improve the car's performance.

DRIVE LE MANS!

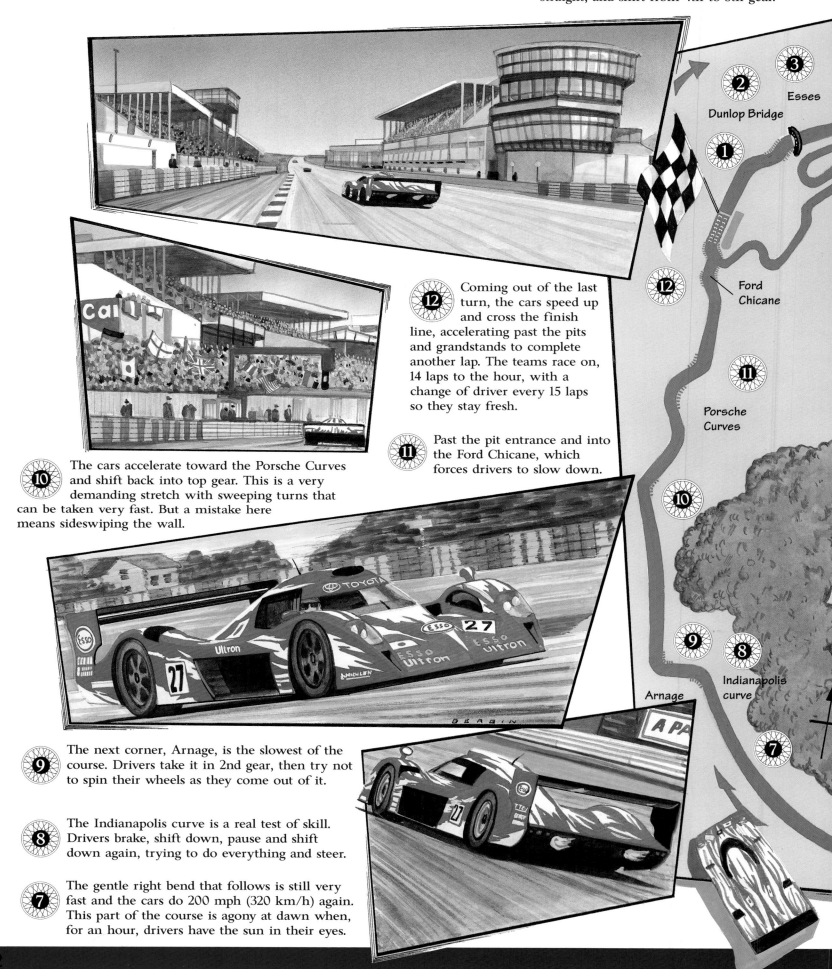

1 Drivers accelerate past the line of pits, garages, and the crowded grandstands along the front straight, and shift from 4th to 5th gear.

12 Coming out of the last turn, the cars speed up and cross the finish line, accelerating past the pits and grandstands to complete another lap. The teams race on, 14 laps to the hour, with a change of driver every 15 laps so they stay fresh.

11 Past the pit entrance and into the Ford Chicane, which forces drivers to slow down.

10 The cars accelerate toward the Porsche Curves and shift back into top gear. This is a very demanding stretch with sweeping turns that can be taken very fast. But a mistake here means sideswiping the wall.

9 The next corner, Arnage, is the slowest of the course. Drivers take it in 2nd gear, then try not to spin their wheels as they come out of it.

8 The Indianapolis curve is a real test of skill. Drivers brake, shift down, pause and shift down again, trying to do everything and steer.

7 The gentle right bend that follows is still very fast and the cars do 200 mph (320 km/h) again. This part of the course is agony at dawn when, for an hour, drivers have the sun in their eyes.

Esses

Dunlop Bridge

Ford Chicane

Porsche Curves

Indianapolis curve

Arnage

A PA

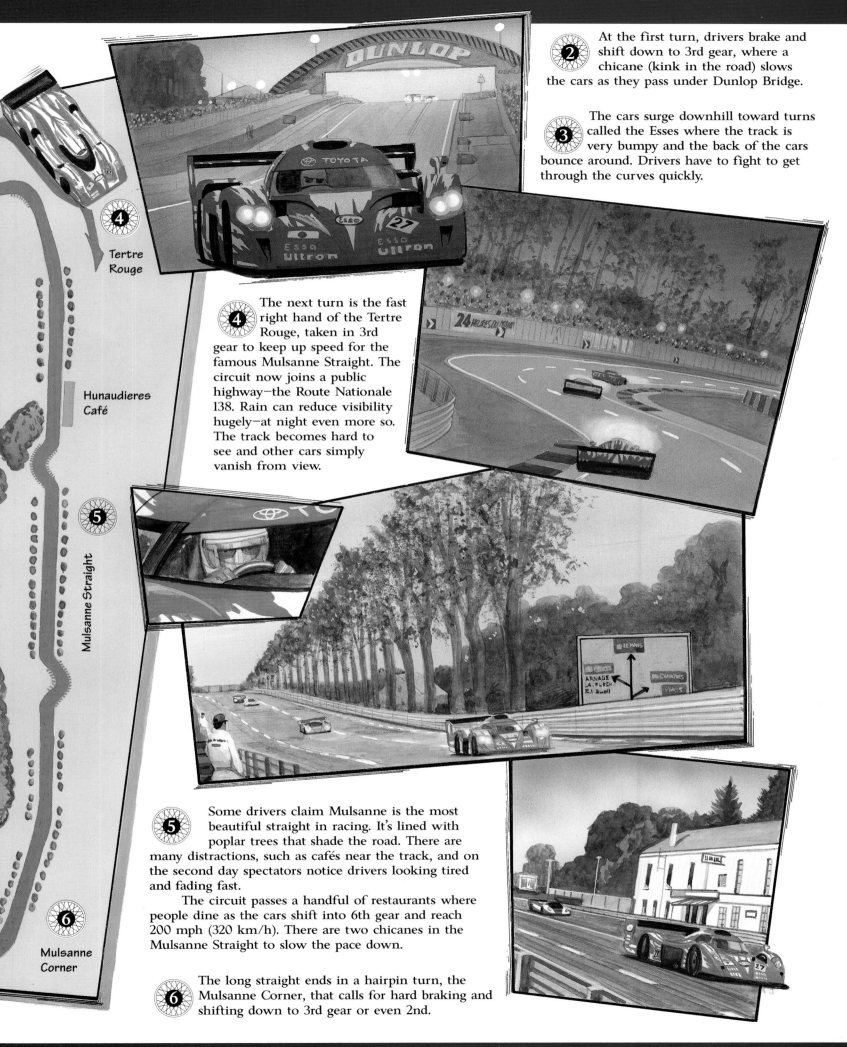

② At the first turn, drivers brake and shift down to 3rd gear, where a chicane (kink in the road) slows the cars as they pass under Dunlop Bridge.

③ The cars surge downhill toward turns called the Esses where the track is very bumpy and the back of the cars bounce around. Drivers have to fight to get through the curves quickly.

Tertre Rouge

④ The next turn is the fast right hand of the Tertre Rouge, taken in 3rd gear to keep up speed for the famous Mulsanne Straight. The circuit now joins a public highway—the Route Nationale 138. Rain can reduce visibility hugely—at night even more so. The track becomes hard to see and other cars simply vanish from view.

Hunaudieres Café

Mulsanne Straight

⑤ Some drivers claim Mulsanne is the most beautiful straight in racing. It's lined with poplar trees that shade the road. There are many distractions, such as cafés near the track, and on the second day spectators notice drivers looking tired and fading fast.

The circuit passes a handful of restaurants where people dine as the cars shift into 6th gear and reach 200 mph (320 km/h). There are two chicanes in the Mulsanne Straight to slow the pace down.

⑥ The long straight ends in a hairpin turn, the Mulsanne Corner, that calls for hard braking and shifting down to 3rd gear or even 2nd.

Mulsanne Corner

GLOSSARY

Acceleration: The amount of time it takes for a car to increase its speed.

Aerofoils: The short, stubby wings fitted front and back on modern racing cars to give them extra grip on the track.

Afterburner: A device that squirts fuel into the hot exhaust gases of a jet engine to give it extra thrust.

Anti-lock brakes (ABS): Brakes that sense when wheels are about to skid and then ease off for a fraction of a second. They stop drivers from sliding out of control.

Carbon fiber: A high-tech material made of fibers woven and mixed with glues. It can be formed into any shape, then heated to make a material harder than steel.

Checkered flag: The end of a race is marked by an official waving a black and white checkered flag.

Chicane: A set of sharp bends flowing first one way then another in the circuit. They slow cars down.

Circuit: A car racing track and all the other parts that go with it—from pits to garages and grandstands.

Cylinders: Metal sockets in an engine where fuel and air burn and in which the pistons slide up and down.

Endurance racing: Long-distance races or races that last for a very long time. High speed is less important here than reliability.

Formula 1: The most prestigious form of motor racing. Today it is a multi-million dollar sport that takes place around the world. The nearest U.S. equivalent is Indy car racing.

Fuel-management system: The electronic controls that regulate how much fuel an engine needs to burn to get the best performance.

Grand Prix: Individual races in different countries that make up the Formula 1 racing year.

Grandstand: The bank of seats at a race track where spectators get the best view— usually near the pits.

▲ Ferrari F 300

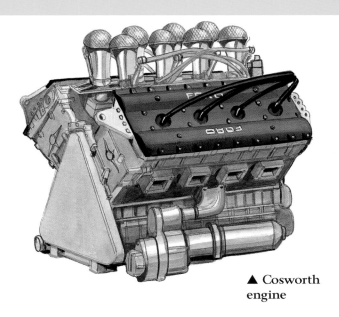

▲ Cosworth engine

Hairpin turn: A very sharp bend that looks like the rounded end of a hairpin.

Horsepower: A unit for measuring an engine's power, based on the amount of effort a horse exerts as it pulls a load.

Lap: A single complete circuit of a track from start to finish. A race is made up of many laps—100 or more is quite common.

Liter: The volume of an engine's cylinders is measured in liters or cubic inches.

Mach: A handy measuring unit for objects traveling faster than the speed of sound. Mach 1 equals the speed of sound.

Piston engine: An engine in which power comes from fuel burned in cylinders. The hot expanding gases move the pistons up and down.

Pits: The places beside the track where crews add fuel, change tires, or make repairs during a race.

Pole position: The best position from which to start a race. It is at the head of the pack.

Radiator: All engines create large amounts of heat they need to get rid of. Usually a radiator placed in the air stream does this.

Revs: The number of times per minute that the pistons of an engine complete their cycle.

Slicks: Completely treadless tires used for racing on dry tracks. They are very wide.

Streamlined: A body shape that is smooth and tapered so it slips through the air with the least wind resistance.

Supercharger: A pump that blows hot exhaust gases back into the engine's cylinders. It allows the engine to burn more fuel and produce a surge of power.

Supersonic: A car or plane that can go faster than the speed of sound.

Top gear: The highest gear, used when a car is going all out.

Wind tunnels: Testing tunnels through which a draft is blown at high speed to discover car (or aircraft) shapes that flow most smoothly through the air.

TIMELINE

▼ Type 35 Bugatti

1895 Emile Levassor wins the first car race between Paris and Bordeaux, France.

1899 The first woman to take part in a car race, Mme. Labrousse, comes in fifth in a race between Paris and Spa.

1900 The first circuit race, the Course du Catalogue, is run in two laps on a triangle of roads at the French town of Melun. Girardot wins in a Panhard.

1906 The first French Grand Prix is run in France near Le Mans. Held over two days, the race's winner is the Hungarian, Ferenc Szisz, in a Renault.

1907 The first car racing track in Britain, Brooklands, is built by HF Locke-King in Surrey. It is notable for its banked curves.

1910 The "fastest man in America," Barney Oldfield, sets a new World Speed Record of 131 mph (210 km/h) in a Blitzen Benz.

▼ Tazio Nuvolari

1911 Ray Harroun wins the first Indianapolis 500 race in a Marmon Wasp. The race has been the pinnacle of U.S. car racing ever since.

1914 The French Grand Prix, held near Lyon, is dominated by Mercedes, who take first, second, and third place.

1924 Ettore Bugatti introduces his Type 35 at a race in Lyon. His car becomes one of the most successful racing cars ever.

1927 Bentley wins the Le Mans 24-hour race. Victories follow in the next three years.

1929 The first Monaco Grand Prix is held. Grover-Williams, driving a Bugatti Type 35, wins in 3 hours, 56 minutes.

1934 Mercedes-Benz and Auto Union cars set new standards in racing car design and technology.

1935 Sir Malcolm Campbell breaks the Land Speed Record twice, eventually raising it to 301 mph (484 km/h).

1935 Tazio Nuvolari beats German teams for a great victory in the German Grand Prix.

1938 John Cobb drives a 27-liter Railton to a Land Speed Record of 350 mph (563 km/h). After World War II he increases it to 634 km/h (394 mph).

1950 The World Driver's Championship is introduced in Grand Prix Racing, with points awarded for victories throughout each season.

1955 The worst accident in car racing history occurs at Le Mans when a Mercedes-Benz 300SLR plows into spectators, killing 83. Mike Hawthorn wins the race in a Jaguar D-type.

▲ Jim Clark

1957 Juan Manuel Fangio wins the German Grand Prix in a Maserati 250F, his fifth and last World Championship.

1960 A transition from front to rear-engine Grand Prix cars occurs. The last victory for a front-engine car—a Ferrari Dino 246 driven by Phil Hill—is the year's Italian Grand Prix winner.

1963 Colin Chapman brings monocoque design to modern racing with his Lotus 25.

1966 The Ford GT40 wins first/second/third place at Le Mans. GT40s win again in 1968 and 1969.

1967 Jim Clark drives a Lotus 49, with a Cosworth V-8 engine, to victory in the Dutch Grand Prix. The Cosworth engine becomes one of the most successful racing engines ever.

1969 Airfoils first appear on Formula 1 cars to create downforce and make the cars hold the track at high speeds.

1976 Niki Lauda survives a horrific accident in the German Grand Prix but resumes racing in six weeks.

1978 "Big Daddy," dragster racer Don Garlits, wows the crowd at Santa Pod, the British drag strip. The Lotus 79 design, with its ground-effects technology, dominates the Formula 1 season. Mario Andretti wins the World Championship.

1983 Richard Noble breaks the Land Speed Record with Thrust 2, which reaches a speed of 633 mph (1,019 km/h) at Black Rock Desert, U.S.A.

1988 Ayrton Senna wins his first World Championship in his turbocharged Honda-powered McLaren. Teammate Alain Prost is second. Between them they win 16 out of 17 races this season.

1992 Al Unser Jr. wins the Indy 500 by the closest-ever margin—.043 of a second.

1993 Alain Prost wins his fourth title in a Renault Williams.

1994 Ayrton Senna dies in a crash at the San Marino Grand Prix. New safety regulations are introduced as a result.

1997 Andy Green drives Thrust SSC to a new Land Speed Record of 763 mph (1,220 km/h).

1998 A Porsche GT1 wins Le Mans after the leading Toyota breaks down, making 16 wins for Porsche in this event.

1998 Mercedes McLaren and Mika Hakkinen win the Formula 1 World Championship after a season-long battle with Ferrari and Michael Schumacher.

▼ Thrust SSC, 1997

INDEX

A B

Alfa Romeo P3 18-19
Barnato, Woolf 15
Bentley 14, 15
Birkin, Tim (Sir Henry) 14, 15
Black Rock Desert, Nevada 35
Blitzen Benz 12-13
Blower Bentley 14-15
Bluebird (1935) 20-21
body shell 40, 41
Bonneville Salt Flats, Utah 21
brakes 41
British Grand Prix 1998 38
Bugatti Type 35 16-17
Burman, Bob 12

C

CAD (computer assisted design) 40
Campbell, Sir Malcolm 20, 21
carbon fiber 40, 41
chain drive 13
Chapman, Colin 26
Clark, Jimmy 26, 27

D

dampers (shock absorbers) 8
Daytona Beach, Florida 20
1909 12
1932 21
drag racing 30, 31
Dutch Grand Prix 1967 26

E

engines
Ford Cosworth DFV 27
on Lotus 49 24
top fuel dragster 30

F

F1 cars, building 40-41
Fangio, Juan Manuel 22, 23
Ferrari F 300 38-39
Ford GT40 28-29
Formula 1 cars see F1 cars
French Grand Prix
1906 8, 9
1914 12
1924 17

G

Garlits, Don 30
German Grand Prix
1935 18
1957 23
Grand Prix 46-47
see also under British, Dutch, French, German, Monaco
Green, Andy 34
Grover-Williams, William 17

H

Harroun, Ray 10, 11
Hawthorn, Mike 24, 25
helmet (Indy Car) 33
Hemery, Victor 12
Hill, Graham 26

I

Ickx, Jacky 29
Indianapolis Motor Speedway 11, 32
Indy 500 race 33
1911 10, 11
1992 32
Indy Car 32-33
Irvine, Eddie 39

J K

Jaguar D-type 24-25
Kidston, Glen 15

L

land speed records 46, 47
1909 12
1924 21
supersonic 34
Le Mans 24-hour endurance race
1927-30 14
1998 36
Le Mans racing circuit
driving 42-43
1906 8, 9
1920s-1930s 15
1955 25
1960s 28, 29
Lotus 49 26-27
Lyons, Sir William 24

M

Mach 1.02 34
Marmon Wasp 10-11
Maserati 250F 22-23
Mercedes Benz 12
Monaco Grand Prix 17
monocoque see body shell

N O

nitro-methanol fuel 30, 31
Noble, Richard 34
nose design 39
Nuvolari, Tazio 18, 19
Oldfield, Barney 12, 13

P Q

pit crews 28, 32
Porsche
911 37
GT1 36-37
Porsche, Ferdinand 36
quick-release rims 8, 9

R

race cars
building 40-41
changes to 38
rear-view mirror 11
Renault (1906) 8-9

S

Schumacher, Michael 38, 39
speeds see land speed records
steering wheels 41
supercharger 15
Szisz, Ferenc 8, 9

T

Thrust 2 34
Thrust SSC 34-35
tire changes 8, 9
tires, slick 31
top-fuel dragster 30-31

U V W

Unser Jr., Al 32
wheel changes 28
wheels, aluminum 34
wind tunnel testing 39, 40

Acknowledgments

Franklin Watts would like to thank the following for their help in creating this book:

Fiat Auto UK
Ford Motor Company
Mercedes Benz (United Kingdom) Limited

The Bentley "B-in-wings" badge is a registered trade mark and is used with the permission of Bentley Motors Limited
Lynn Bresler for the index